ROBOT WO

ROBOTS IN SPACE

by Jenny Fretland VanVoorst

Ideas for Parents and Teachers

Pogo Books let children practice reading informational text while introducing them to nonfiction features such as headings, labels, sidebars, maps, and diagrams, as well as a table of contents, glossary, and index.

Carefully leveled text with a strong photo match offers early fluent readers the support they need to succeed.

Before Reading

- "Walk" through the book and point out the various nonfiction features. Ask the student what purpose each feature serves.
- Look at the glossary together. Read and discuss the words.

Read the Book

- Have the child read the book independently.
- Invite him or her to list questions that arise from reading.

After Reading

- Discuss the child's questions. Talk about how he or she might find answers to those questions.
- Prompt the child to think more. Ask: Can you think of another environment where robots could go more safely than people? What other dangerous jobs do you think robots could be used for?

To Liam—JFV

Pogo Books are published by Jump!
5357 Penn Avenue South
Minneapolis, MN 55419
www.jumplibrary.com

Copyright © 2016 Jump!
International copyright reserved in all countries. No part of this book may be reproduced in any form without written permission from the publisher.

Library of Congress Cataloging-in-Publication Data

Fretland VanVoorst, Jenny, 1972-
  Robots in space / by Jenny Fretland VanVoorst.
    pages cm. – (Robot world)
  Includes bibliographical references and index.
  ISBN 978-1-62031-219-3 (hardcover: alk. paper) –
  ISBN 978-1-62031-424-1 (paperback) –
  ISBN 978-1-62496-306-3 (ebook)
  1. Robots–Juvenile literature.
  2. Space robotics–Juvenile literature. I. Title.
  TL1097.F745 2015
  629.43–dc23

2015020999

Series Designer: Anna Peterson
Photo Researchers: Anna Peterson and Michelle Sonnek

Photo Credits: Alamy, 18-19; Corbis, 1, 14-15, 16; NASA, 6-7, 10-11, 12-13, 17, 20-21, 23; Science Source, cover, 3; Shutterstock, 4, 5, 8, 9.

Printed in the United States of America at Corporate Graphics in North Mankato, Minnesota.

TABLE OF CONTENTS

CHAPTER 1
What Do They Do? 4

CHAPTER 2
How Do They Work? 8

CHAPTER 3
Meet the Robots 16

ACTIVITIES & TOOLS
Try This! 22
Glossary 23
Index 24
To Learn More 24

CHAPTER 1
WHAT DO THEY DO?

Have you ever wanted to visit Mars? Explore the moon? It sounds exciting. But it is also very dangerous. There's no air for us to breathe. Food and water are nowhere to be found.

But robots don't need air. They don't need food or water. They just need a source of power and a set of instructions.

CHAPTER 1   5

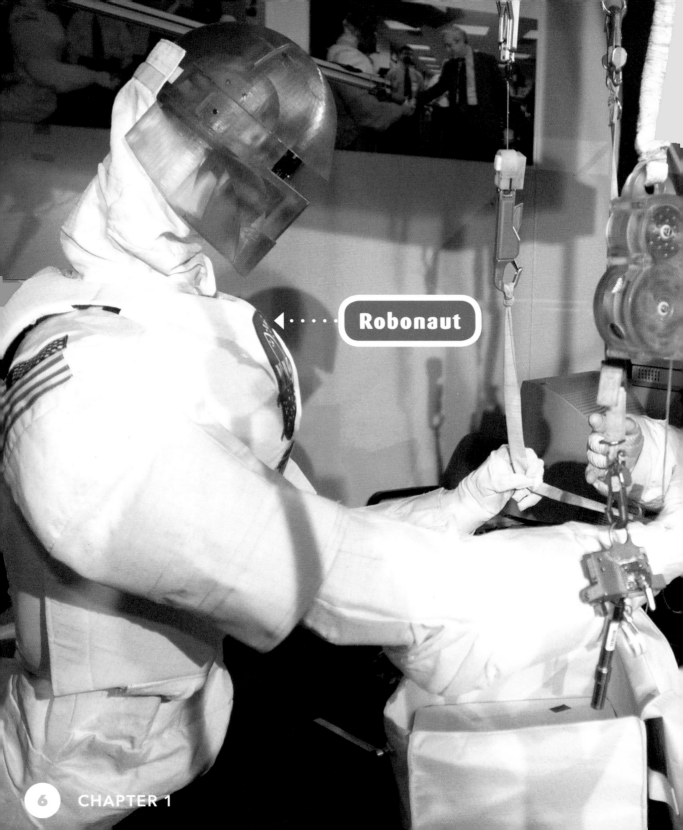

Space robots orbit Earth. They work inside the **International Space Station**. They explore other planets. Some space robots serve as human helpers. Others do things that people can't—at least not right now.

DID YOU KNOW?

Robots work off of **programs**. A program is a set of directions. It tells a robot what to do in great detail. It leaves no room for error.

1 ....
2 ....
3 ....
4 ....
5 ....

CHAPTER 2
HOW DO THEY WORK?

Robots that work as human helpers are often remote controlled.

Other robots work independently. Robots on a distant planet can't depend on people for directions. They are too far away.

CHAPTER 2

As a result, many space robots are **autonomous**. They gather **data** about their surroundings. Then they use that data to decide how to act.

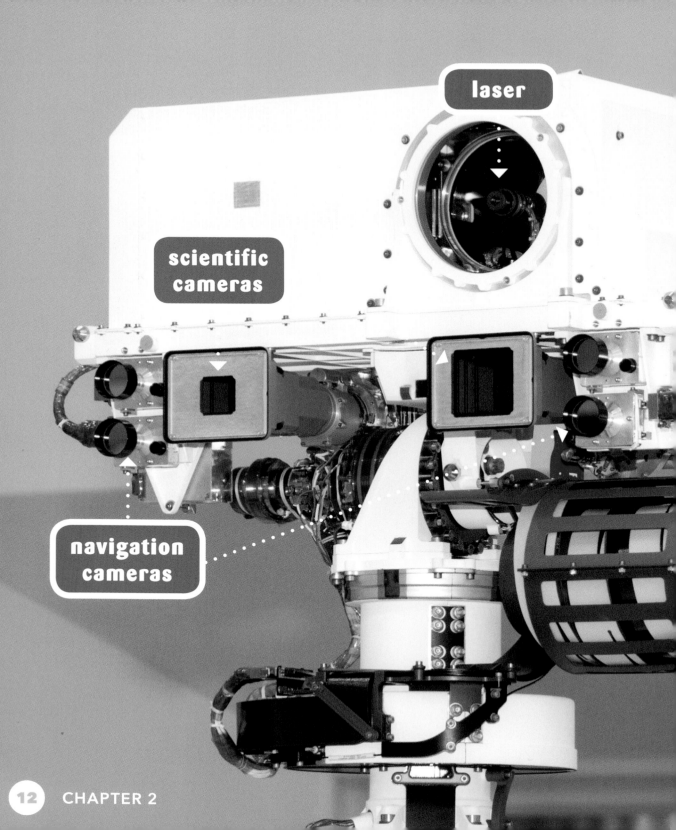

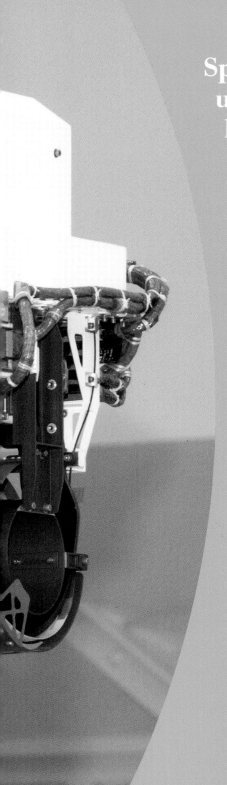

Space robots gather data using **sensors**. They might have cameras and **GPS**. They might have **lidar** or **radar**. They may have a hand that senses pressure.

CHAPTER 2

If a rock is in a robot's path, the robot will use its sensors to spot what else is around it. Then its computer uses that information to find a safe path.

> **TAKE A LOOK!**
>
> Robots sense what's around them. They "think" about the next step. Then they take action. This is called the sense-think-act cycle. Not all robots are capable of all three steps. But many space robots are.
>
> **1. SENSE**
>
> **2. THINK**
>
> **3. ACT**

CHAPTER 2

CHAPTER 3
MEET THE ROBOTS

We build robots to do jobs that a person would otherwise do. That's why many robots are based on the human form. For example, Robonaut looks like a human **torso**.

This space robot has hands and arms. It uses the same tools that humans use. It helps scientists on the International Space Station.

CHAPTER 3   17

Curiosity is a Mars **rover**. It is an autonomous robot. It has a program. But it senses, thinks, and acts to carry it out.

DID YOU KNOW?

Curiosity is studying the rocks and climate on Mars. The data it gathers will help scientists learn whether Mars could have ever supported life.

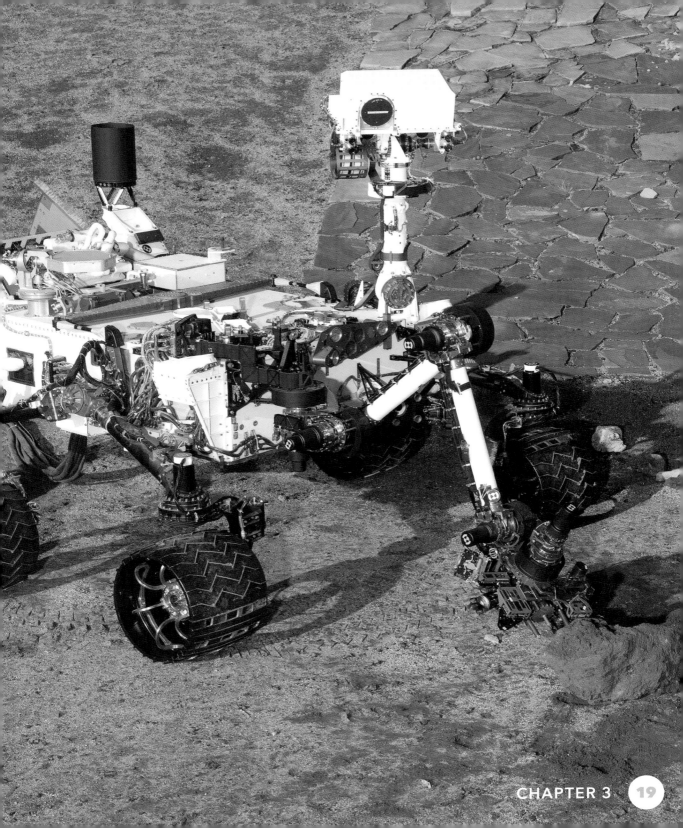

CHAPTER 3

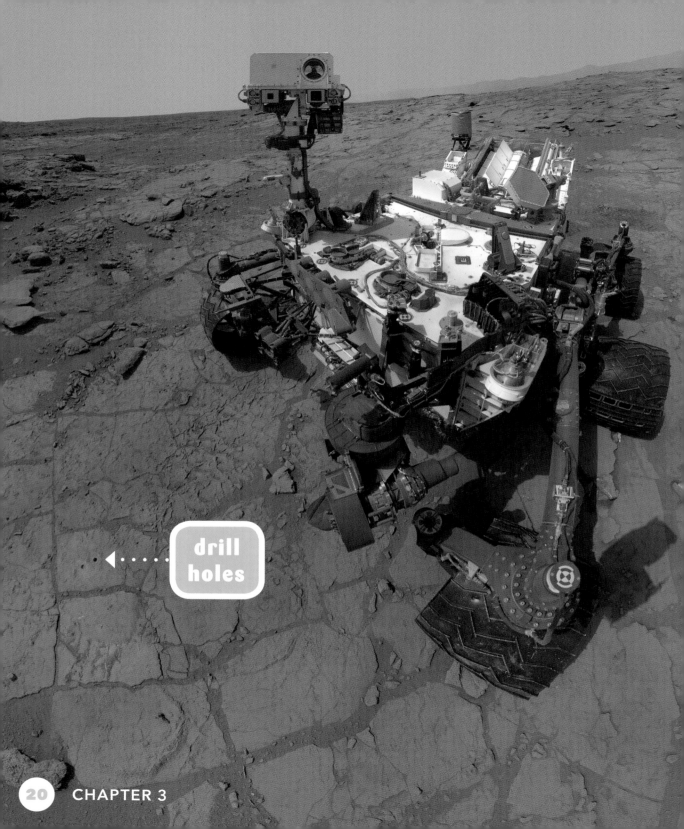

Curiosity collects data. It does experiments. Because of it, we know things about Mars we otherwise never could.

Where do you think robots will go next? What do you think they will teach us?

CHAPTER 3

ACTIVITIES & TOOLS

TRY THIS!

ROBOT SENSING

Senses are very important for robots. Try this experiment to learn what it takes to get a robot to "think" and act.

- blindfold
- items to create an obstacle course (chairs, pillows, cushions, boxes)

① **Set up an obstacle course in a safe place in your home or yard. Create simple, narrow paths with some twists and turns.**

② **Blindfold a friend and bring her to the obstacle course. Explain to her that she's a remote-controlled robot and that you are the controller. For her to get through the course, she has to use her sense of hearing to listen to your instructions. She also needs to use her sense of touch to be aware of obstacles in her path.**

③ **Give very specific instructions for your robot to follow — literally step by step.**

④ **After your friend finishes the course, talk about what instructions were helpful and which were not. Then let your friend recreate the obstacle course and have you follow her commands. Compare experiences as both robot and controller.**

GLOSSARY

autonomous: Able to make certain decisions without human input.

data: Facts about something that can be used in calculating, reasoning, or planning.

GPS: A navigation system that uses satellite signals to find the location of a radio receiver on or above the earth's surface; abbreviation of global positioning system.

International Space Station: A structure orbiting Earth where scientists live and work, performing experiments in space.

lidar: A device that uses laser beams to detect and locate objects.

program: A set of instructions that a robot follows.

radar: A device that uses radio waves to detect and locate objects.

rover: A kind of mobile exploratory robot designed to move over rough ground.

sensors: Onboard tools that serve as a robot's eyes, ears, and other sense organs so that the robot can create a picture of the environment in which it operates.

torso: A person's upper body, from the waist up.

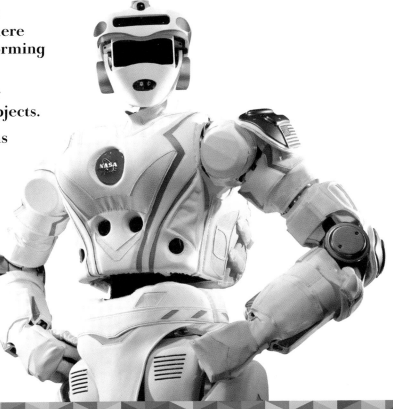

ACTIVITIES & TOOLS

INDEX

acting 11, 14, 18
arms 17
autonomous robots 11, 18
cameras 13
computer 14
Curiosity 18, 21
data 11, 13, 18, 21
Earth 7
experiments 21
GPS 13
hands 13, 17
International Space station 7, 17
lidar 13
Mars 4, 18, 21
moon 4
orbit 7
planets 7, 9
pressure sensors 13
program 7, 18
radar 13
remote control 8
sense-think-act cycle 14, 18
sensors 13, 14
tools 17

TO LEARN MORE

Learning more is as easy as 1, 2, 3.
1) Go to www.factsurfer.com
2) Enter "robotsinspace" into the search box.
3) Click the "Surf" to see a list of websites.

With factsurfer, finding more information is just a click away.